For My Off-Planet Benefactors

Who Open My Mind and Heart

ALL OF CREATION IS CURVES, WAVES & OSCILLATIONS

TABLE OF CONTENTS

INTRODUCTION 1

- THE ROLE OF STRAIGHT LINES IN MATHEMATICS AND DAILY LIFE 1
- CHALLENGING THE STRAIGHT LINE: INSIGHTS FROM NON-EUCLIDEAN GEOMETRY 2
- THE IMPACT OF GENERAL RELATIVITY ON OUR UNDERSTANDING OF SPACE 3
- QUANTUM MECHANICS AND THE BREAKDOWN OF CLASSICAL GEOMETRY 3
- THE INFLUENCE OF CELESTIAL MOVEMENTS ON PERCEIVED STRAIGHT LINES 4
- PHILOSOPHICAL PERSPECTIVES: MAYA AND LEELA 4
- THE WAKING STATE (JAGRAT) AND THE ILLUSION OF STRAIGHT LINES 5
- INTEGRATING PERSPECTIVES: A UNIFIED UNDERSTANDING 6

THE ILLUSION OF LINEARITY 7

- THE HISTORICAL SIGNIFICANCE OF STRAIGHT LINES 8
- PRACTICAL USES OF STRAIGHT LINES 8
- THE LIMITATIONS OF EUCLIDEAN GEOMETRY 9
- THE QUANTUM PERSPECTIVE 11
- PHILOSOPHICAL IMPLICATIONS: MAYA AND LEELA 12
- THE WAKING STATE (JAGRAT) AND THE ILLUSION OF STRAIGHT LINES 12

BEYOND EUCLID - THE BIRTH OF NON-EUCLIDEAN GEOMETRY 14

- THE FOUNDATIONS OF EUCLIDEAN GEOMETRY 14
- HYPERBOLIC GEOMETRY 17
- ELLIPTIC GEOMETRY 18
- THE IMPACT OF NON-EUCLIDEAN GEOMETRY 18
- VISUALIZING NON-EUCLIDEAN GEOMETRY 19

EINSTEIN'S CURVED SPACE 22

- THE BIRTH OF GENERAL RELATIVITY 22
- SPACETIME AND CURVATURE 23
- GEODESICS: THE NEW STRAIGHT LINES 23
- THE WARPING OF SPACETIME BY GRAVITY 24
- EXPERIMENTAL EVIDENCE FOR GENERAL RELATIVITY 24
- BLACK HOLES: EXTREME CURVATURE OF SPACETIME 25
- GRAVITATIONAL WAVES: RIPPLES IN SPACETIME 26
- THE IMPLICATIONS FOR STRAIGHT LINES 26
- VISUALIZING CURVED SPACETIME 27

THE QUANTUM WORLD 29

- THE BIRTH OF QUANTUM MECHANICS 29
- WAVE-PARTICLE DUALITY 30
- THE WAVE FUNCTION AND PROBABILITY DISTRIBUTIONS 30
- HEISENBERG'S UNCERTAINTY PRINCIPLE 31
- THE BREAKDOWN OF CLASSICAL TRAJECTORIES 32

Other Works By This Author:

Weapons Grade Moxie: Unleashing Your Inner Strength to Conquer Fear and Toxic People

Wholeness Through Illness- Finding Meaning, Healing, and Grace

You Are Infinite!: Co-Creating the Fractal Holograph

Unstuck: Experiencing the Anahata Nad

The Pineal Portal: Unlocking The Secrets of The Third Eye

Life In The Bliss Lane: A Guide To Wellness, Self-Love, And Joy

Killing Time: Breaking Free From Temporal Chains

Level Up With Gratitude: The Ultimate Bio-Hack For Happiness

Ardhanarishwara Charitra: The Metaphysical Wisdom of Gender Fluidity

Vasudhaiva Kutumbakam: The Limitless Power of Our Desi Roots

The Atma's Journey: Tarot Wisdom Through the Ramayana & Mahabharata

You Make Me Sick: Virtue Signaling & Narcissistic Abuse

Toxic Sibling Estrangement: Reclaiming Your Inner Peace

Hidden Branches In The Family Tree: Navigating the NPE Experience

Beyond Binary: An Exploration in Gender and Sexuality

Pharmajuana: Guide to Cannabis for Cancer

Why Straight Lines Don't Exist: Exploring Geometric Truths

How to Gym: Becoming Fitness Itself

P.E.A.K.: Phenomenological Experiences At Kailash

@2024

LIFE IN THE BLISS LANE

DR. SUNAYANA SHIVANGI PANDÉ

- Quantum Tunneling 32
- Implications for Geometry and Straight Lines 33
- Quantum Entanglement and Non-Locality 33
- Visualizing Quantum Paths 34

THE SPINNING UNIVERSE 36

- Earth's Rotation and Revolution 36
- The Sun's Motion Through the Milky Way 38
- The Combined Motions 38
- The Implications for Perceived Straight Lines 39
- Visualizing the Spinning Universe 40

MAYA AND LEELA - PHILOSOPHICAL PERSPECTIVES 42

- The Concept of Maya in Advaita Vedanta 43
- Leela: The Divine Play of the Universe 44
- Maya and Leela: Complementary Perspectives 46
- Implications for Geometric Constructs 47

JAGRAT - THE WAKING STATE 49

- The Role of Maya in Jagrat 49
- Tools of Communication and Understanding 50
- Inherent Limitations of Perception 51
- The Analogy of the Airport Lounge 52
- Embracing the Inherent Limitations 54

INTEGRATING PERSPECTIVES 57

- Mathematics and Idealization 57
- Science and Empirical Reality 59
- Metaphysics and Ultimate Reality 61
- The Unified Understanding 62

PRACTICAL IMPLICATIONS 65

- Science and Engineering 65
- Philosophy and Metaphysics 67
- Art and Design 68

THOUGHT EXPERIMENTS AND EXERCISES 71

- 1. Draw a Line in Motion 71
- 2. Visualize Geodesics 72
- 3. Simulate Curved Space 73
- 4. Explore Quantum Paths 74

CONCLUSION: EMBRACING CURVED REALITIES 77

- Final Thoughts 77
- Embracing a New Paradigm 79

PRACTICAL IMPLICATIONS 79
CONTINUING THE JOURNEY 80
ABOUT THE AUTHOR **81**

WHY STRAIGHT LINES DON'T EXIST: EXPLORING GEOMETRIC TRUTHS

In the world of mathematics, science, and daily life, straight lines are fundamental. They shape our understanding of space, guide our constructions, and define the shortest paths between points. From the ancient geometries of Euclid to modern engineering marvels, straight lines have been considered the bedrock of our spatial understanding. However, this reliance on straight lines is built on an illusion. Upon closer examination, through the lenses of mathematics, science, and metaphysics, it becomes clear that straight lines do not exist in reality. Instead, they are constructs of human perception, useful for navigating our immediate experiences but ultimately part of the grand illusion of Maya and Leela.

INTRODUCTION

THE ROLE OF STRAIGHT LINES IN MATHEMATICS AND DAILY LIFE

In Euclidean geometry, a straight line is the shortest distance between two points. This concept has been foundational in mathematics for over two millennia. It underpins the way we measure distances, create maps, and design structures. Architects use straight lines to plan buildings, engineers use them to design machinery, and artists use them to create perspective in their work. The simplicity and practicality of straight lines make them indispensable tools in our everyday lives.

WHY STRAIGHT LINES DON'T EXIST: EXPLORING GEOMETRIC TRUTHS

However, our reliance on straight lines goes beyond their practical utility. They also shape the way we think about space and movement. When we navigate from one place to another, we often think in terms of straight paths, even if the actual journey involves curves and detours. Our mental maps of the world are filled with straight lines that connect different points of interest. This way of thinking is so deeply ingrained that it can be difficult to imagine any other way to conceptualize space.

CHALLENGING THE STRAIGHT LINE: INSIGHTS FROM NON-EUCLIDEAN GEOMETRY

In the 19th century, mathematicians such as Carl Friedrich Gauss, Nikolai Lobachevsky, and János Bolyai began to challenge the Euclidean notion of straight lines. They developed non-Euclidean geometries, which included hyperbolic and elliptic spaces where the traditional rules of straight lines no longer applied.

In hyperbolic geometry, space is curved negatively, like a saddle. The sum of the angles of a triangle is less than 180 degrees, and parallel lines can diverge. In elliptic geometry, space is curved positively, like a sphere. The sum of the angles of a triangle is more than 180 degrees, and there are no true parallel lines; all lines eventually intersect. These geometries reveal that the concept of a straight line is not universal but depends on the underlying geometry of the space.

WHY STRAIGHT LINES DON'T EXIST: EXPLORING GEOMETRIC TRUTHS

THE IMPACT OF GENERAL RELATIVITY ON OUR UNDERSTANDING OF SPACE

Albert Einstein's theory of general relativity further revolutionized our understanding of space and time. General relativity describes gravity as the curvature of spacetime caused by massive objects. According to this theory, the presence of mass and energy warps the fabric of spacetime, causing paths that would be straight in a flat space to curve.

In the context of general relativity, the closest equivalent to a straight line is a geodesic. A geodesic is the path that a free-falling object follows under the influence of gravity. These paths are curved by the presence of mass and energy, demonstrating that straight lines do not exist in the curved spacetime of our universe.

QUANTUM MECHANICS AND THE BREAKDOWN OF CLASSICAL GEOMETRY

At the quantum level, the behavior of particles further disrupts the idea of straight lines. Quantum mechanics reveals that particles do not follow precise paths. Instead, their positions and momenta are described by probability distributions.

Heisenberg's Uncertainty Principle states that we cannot simultaneously know the exact position and momentum of a particle. Consequently, the concept of a well-defined, straight-line trajectory loses its meaning at microscopic scales. The probabilistic nature of particle paths challenges the classical notion of straight lines, introducing a new layer of complexity to our understanding of geometry.

THE INFLUENCE OF CELESTIAL MOVEMENTS ON PERCEIVED STRAIGHT LINES

The rotation and revolution of celestial bodies add another layer to the understanding of why straight lines do not exist in reality. The Earth rotates on its axis approximately once every 24 hours and revolves around the Sun once every 365.25 days. This means that any point on the Earth's surface is constantly moving in a circular path due to rotation and an elliptical path due to revolution.

Additionally, the Sun itself orbits the center of the Milky Way galaxy, moving at an average speed of about 220 kilometers per second (490,000 miles per hour). The entire solar system, including the Earth, is moving through the galaxy. These combined motions mean that any attempt to draw a straight line is influenced by multiple layers of movement, each adding its own curvature to the perceived straightness.

PHILOSOPHICAL PERSPECTIVES: MAYA AND LEELA

In Indian philosophy, the concepts of Maya (illusion) and Leela (divine play) provide profound insights into the nature of reality. Maya, according to Advaita Vedanta, is the cosmic illusion that veils the true nature of Brahman, the ultimate reality. It creates the appearance of a diverse, tangible world where distinctions and forms seem real but are ultimately illusory. Straight

WHY STRAIGHT LINES DON'T EXIST: EXPLORING GEOMETRIC TRUTHS

lines, as part of this perceived world, are thus illusions within the grand illusion of Maya.

Leela, in Vaishnavism, refers to the divine play of the universe, where the material world and its phenomena are seen as the playful expressions of the divine will. The perceived order, including geometric constructs like straight lines, are manifestations of this divine play, lacking independent reality.

THE WAKING STATE (JAGRAT) AND THE ILLUSION OF STRAIGHT LINES

The Mandukya Upanishad describes four states of consciousness: Jagrat (waking state), Swapna (dream state), Sushupti (deep sleep state), and Turiya (the transcendental state). In the waking state (Jagrat), we perceive the world through our senses and engage in activities. This state is characterized by the apparent solidity and distinctness of forms, including the perception of straight lines. The Jagrat state is dominated by the influence of Maya, the cosmic illusion that shapes our perception of reality.

An intriguing analogy compares the waking state (Jagrat) to an airport lounge. In this metaphor, the airport lounge is a transient space where travelers from different destinations interact and prepare for their journeys. Similarly, the waking state is where individuals engage with the world and each other, navigating the transient, illusory nature of physical reality. In this state, straight lines and other geometric constructs are necessary tools for communication and understanding.

WHY STRAIGHT LINES DON'T EXIST: EXPLORING GEOMETRIC TRUTHS

INTEGRATING PERSPECTIVES: A UNIFIED UNDERSTANDING

By integrating the insights from mathematics, science, and metaphysics, we arrive at a unified understanding that straight lines are constructs of the mind rather than features of the physical world. This book will explore these perspectives in detail, offering a comprehensive journey through the geometric truths of our universe. We will challenge traditional notions of geometry, delve into the complexities of curved spaces, and reveal the deeper nature of reality beyond the illusions of straight lines.

∴

As we embark on this journey, we invite you to rethink mathematics and geometry completely. By examining the myth of the straight line, we can gain a deeper appreciation for the complexity and beauty of the universe. We will recognize that our perceptions are shaped by underlying principles that transcend simple geometric constructs, inviting us to explore the true, boundless nature of reality beyond the illusions of form and structure. Through this exploration, we will uncover the profound mysteries that lie beyond the straight lines we once took for granted.

WHY STRAIGHT LINES DON'T EXIST: EXPLORING GEOMETRIC TRUTHS

Straight lines have long been considered the simplest and most fundamental geometric construct. Euclidean geometry, developed by the ancient Greek mathematician Euclid, defines a straight line as the shortest distance between two points. This concept has been foundational in mathematics, influencing fields such as engineering, architecture, and navigation. However, Euclidean geometry operates under the assumption of a flat, infinite plane, which does not hold true in the actual physical universe.

THE ILLUSION OF LINEARITY

In practical terms, straight lines are incredibly useful. They allow us to measure distances, create maps, and construct buildings with precision. Despite their practicality, straight lines are idealized abstractions that simplify the more complex, curved nature of space. The real world, influenced by the curvature of spacetime and the dynamic movements of celestial bodies, does not conform to the simplistic notions of Euclidean geometry.

This chapter explores the historical significance of straight lines in mathematics and geometry, highlighting their practical uses and inherent limitations. It sets the stage for understanding why straight lines are an illusion and introduces the journey to uncover the deeper geometric truths of our universe.

WHY STRAIGHT LINES DON'T EXIST: EXPLORING GEOMETRIC TRUTHS

THE HISTORICAL SIGNIFICANCE OF STRAIGHT LINES

Straight lines have been a cornerstone of geometric thought since the time of the ancient Greeks. Euclid's "Elements," written around 300 BCE, is one of the most influential works in the history of mathematics. In this treatise, Euclid lays out the principles of what we now call Euclidean geometry, including the definition of a straight line as "a line that lies evenly with the points on itself." This definition is intuitive and forms the basis for many geometric constructions.

Euclidean geometry is built on five postulates, which are assumed to be self-evident truths. The first postulate states that a straight line can be drawn between any two points. This postulate underpins the entire structure of Euclidean geometry and has been accepted without question for centuries.

PRACTICAL USES OF STRAIGHT LINES

In practical terms, straight lines are indispensable. They provide the most efficient way to measure distances, as the shortest path between two points is a straight line. This principle is used in various fields:

1. **Engineering**: Engineers use straight lines to design and construct buildings, bridges, and machinery. The precision and simplicity of straight lines allow for accurate calculations and stable structures.

2. **Architecture**: Architects rely on straight lines to create blueprints and plans for buildings. The clarity and predictability of straight lines ensure that structures are safe and aesthetically pleasing.

WHY STRAIGHT LINES DON'T EXIST: EXPLORING GEOMETRIC TRUTHS

3. **Navigation**: Navigators use straight lines to plot courses on maps. Whether by sea, air, or land, the principle of the shortest path is crucial for efficient travel.

4. **Art**: Artists use straight lines to create perspective and depth in their works. The use of vanishing points and horizon lines in drawings and paintings relies on the concept of straight lines to convey three-dimensional space on a two-dimensional surface.

THE LIMITATIONS OF EUCLIDEAN GEOMETRY

While Euclidean geometry is practical and useful, it is based on idealized assumptions that do not hold true in the real world. Euclidean geometry assumes a flat, infinite plane where parallel lines never meet, and the angles in a triangle always add up to 180 degrees. These assumptions are useful for many everyday applications but break down when we consider the true nature of space.

THE CURVED NATURE OF SPACE

In reality, space is not flat. The presence of mass and energy warps space, creating curves and distortions. This curvature is described by the theory of general relativity, developed by Albert Einstein in the early 20th century. According to general relativity, gravity is not a force that acts at a distance but a result of the curvature of spacetime caused by mass and energy.

In this curved spacetime, the shortest distance between two points is no longer a straight line in the Euclidean sense but a curved path called a

WHY STRAIGHT LINES DON'T EXIST: EXPLORING GEOMETRIC TRUTHS

geodesic. Geodesics are the generalization of straight lines to curved spaces. They represent the paths that objects follow when moving under the influence of gravity alone.

THE INFLUENCE OF CELESTIAL BODIES

The movements of celestial bodies add another layer of complexity to our understanding of straight lines. The Earth rotates on its axis and revolves around the Sun, while the Sun itself orbits the center of the Milky Way galaxy. These motions mean that any point on the Earth's surface is constantly moving, and any line drawn on the Earth is part of a dynamic, curved trajectory through space.

Even if we draw what appears to be a straight line on a piece of paper, that line is subject to the rotation and revolution of the Earth, as well as the motion of the solar system through the galaxy. Therefore, the line is not truly straight but follows a complex, spiraling path through space.

NON-EUCLIDEAN GEOMETRY

The development of non-Euclidean geometries in the 19th century provided new ways to understand space and challenged the notion of straight lines. Mathematicians such as Carl Friedrich Gauss, Nikolai Lobachevsky, and János Bolyai independently developed geometries that did not adhere to Euclid's fifth postulate, which states that parallel lines never meet.

In hyperbolic geometry, for example, space is negatively curved, like a saddle. In this geometry, the angles of a triangle add up to less than 180 degrees, and parallel lines can diverge. In elliptic geometry, space is positively curved, like a sphere. In this geometry, the angles of a triangle add up to more than 180 degrees, and there are no true parallel lines; all lines eventually intersect.

These non-Euclidean geometries reveal that the concept of a straight line is not universal but depends on the underlying geometry of the space. What we perceive as straight lines in Euclidean geometry are actually curved in these non-Euclidean spaces.

THE QUANTUM PERSPECTIVE

At the quantum level, the behavior of particles further disrupts the idea of straight lines. Quantum mechanics, the branch of physics that deals with the behavior of particles on the smallest scales, reveals that particles do not follow precise paths. Instead, their positions and momenta are described by probability distributions.

Heisenberg's Uncertainty Principle, a fundamental concept in quantum mechanics, states that we cannot simultaneously know the exact position and momentum of a particle. This principle means that the concept of a well-defined, straight-line trajectory loses its meaning at microscopic scales. The paths of particles are inherently uncertain and probabilistic, challenging the classical notion of straight lines.

PHILOSOPHICAL IMPLICATIONS: MAYA AND LEELA

In Indian philosophy, the concepts of Maya (illusion) and Leela (divine play) offer profound insights into the nature of reality. Maya, according to Advaita Vedanta, is the cosmic illusion that veils the true nature of Brahman, the ultimate reality. It creates the appearance of a diverse, tangible world where distinctions and forms seem real but are ultimately illusory. Straight lines, as part of this perceived world, are thus illusions within the grand illusion of Maya.

Leela, in Vaishnavism, refers to the divine play of the universe, where the material world and its phenomena are seen as the playful expressions of the divine will. The perceived order, including geometric constructs like straight lines, are manifestations of this divine play, lacking independent reality.

THE WAKING STATE (JAGRAT) AND THE ILLUSION OF STRAIGHT LINES

The Mandukya Upanishad describes four states of consciousness: Jagrat (waking state), Swapna (dream state), Sushupti (deep sleep state), and Turiya (the transcendental state). In the waking state (Jagrat), we perceive the world through our senses and engage in activities. This state is characterized by the apparent solidity and distinctness of forms, including the perception of straight lines. The Jagrat state is dominated by the influence of Maya, the cosmic illusion that shapes our perception of reality.

WHY STRAIGHT LINES DON'T EXIST: EXPLORING GEOMETRIC TRUTHS

An intriguing analogy compares the waking state (Jagrat) to an airport lounge. In this metaphor, the airport lounge is a transient space where travelers from different destinations interact and prepare for their journeys. Similarly, the waking state is where individuals engage with the world and each other, navigating the transient, illusory nature of physical reality. In this state, straight lines and other geometric constructs are necessary tools for communication and understanding.

∴

Straight lines, while fundamental to our understanding and navigation of the world, are ultimately tools of perception, part of the grand illusions of Maya and Leela. Mathematics, science, and metaphysics all converge on this realization, each offering unique insights into why straight lines do not exist in reality but are constructs of the mind.

As we move forward in this book, we will delve deeper into the various perspectives that challenge the notion of straight lines. We will explore the implications of non-Euclidean geometries, the curved nature of space in general relativity, the probabilistic paths of particles in quantum mechanics, and the philosophical insights of Maya and Leela. Through this journey, we will uncover the deeper geometric truths of our universe, inviting you to rethink mathematics and geometry completely.

WHY STRAIGHT LINES DON'T EXIST: EXPLORING GEOMETRIC TRUTHS

In the 19th century, the field of mathematics underwent a revolution with the development of non-Euclidean geometries. Mathematicians such as Carl Friedrich Gauss, Nikolai Lobachevsky, and János Bolyai began to challenge the established notions of Euclidean geometry, which had been the standard for over two millennia. These new geometries—hyperbolic and elliptic—revealed that the traditional rules of straight lines no longer applied universally. This chapter delves into the development of non-Euclidean geometries and how they fundamentally challenge the concept of straight lines.

BEYOND EUCLID - THE BIRTH OF NON-EUCLIDEAN GEOMETRY

THE FOUNDATIONS OF EUCLIDEAN GEOMETRY

Euclidean geometry, formulated by the ancient Greek mathematician Euclid in his seminal work "Elements," is based on a set of axioms and postulates that define the nature of points, lines, and planes. One of the key postulates, known as the parallel postulate, states that given a line and a point not on that line, there is exactly one line through the point that is parallel to the original line. This postulate underpins the entire structure of Euclidean geometry and leads to the familiar properties of shapes and angles.

WHY STRAIGHT LINES DON'T EXIST: EXPLORING GEOMETRIC TRUTHS

For centuries, mathematicians attempted to prove the parallel postulate using Euclid's other axioms but were unsuccessful. This led to the realization that alternative geometries could exist where the parallel postulate did not hold, paving the way for the development of non-Euclidean geometries.

CARL FRIEDRICH GAUSS: THE PIONEER OF NON-EUCLIDEAN GEOMETRY

Carl Friedrich Gauss, one of the greatest mathematicians of all time, was among the first to consider the possibility of non-Euclidean geometry. Although he did not publish his work on the subject, his correspondence with other mathematicians reveals that he had explored the idea extensively. Gauss's insights laid the groundwork for later developments in the field.

Gauss considered the implications of a geometry where the parallel postulate was not valid. He explored the properties of triangles on curved surfaces, such as spheres and hyperbolic planes, and realized that the sum of the angles of a triangle could differ from 180 degrees. His work showed that alternative geometries were not only possible but also consistent and logically sound.

WHY STRAIGHT LINES DON'T EXIST: EXPLORING GEOMETRIC TRUTHS

NIKOLAI LOBACHEVSKY: THE FIRST TO PUBLISH

Nikolai Lobachevsky, a Russian mathematician, was the first to publish a comprehensive account of non-Euclidean geometry. In 1829, he published his work on what is now known as hyperbolic geometry. Lobachevsky's bold assertion was that the parallel postulate could be replaced with an alternative postulate: through a given point not on a line, there are infinitely many lines that do not intersect the original line. This new geometry described a space that is curved negatively, like a saddle.

In hyperbolic geometry, the sum of the angles of a triangle is less than 180 degrees, and parallel lines can diverge. Lobachevsky's work demonstrated that the properties of shapes and angles in this geometry were consistent and coherent, even though they differed from those of Euclidean geometry.

JÁNOS BOLYAI: INDEPENDENT DISCOVERY

Around the same time as Lobachevsky, the Hungarian mathematician János Bolyai independently developed a similar form of non-Euclidean geometry. Bolyai's work, published in 1832, described the same hyperbolic geometry that Lobachevsky had explored. Bolyai's father, Farkas Bolyai, had long been interested in the parallel postulate and had warned his son against attempting to prove it. Despite this, János Bolyai pursued his own ideas and arrived at the same revolutionary conclusions as Lobachevsky.

Bolyai's work included detailed discussions of the properties of hyperbolic space, further solidifying the validity of non-Euclidean geometry. His contributions, alongside those of Lobachevsky, marked a significant turning point in the history of mathematics.

WHY STRAIGHT LINES DON'T EXIST: EXPLORING GEOMETRIC TRUTHS

HYPERBOLIC GEOMETRY

Hyperbolic geometry describes a space that is curved negatively, like a saddle. In this geometry, the familiar rules of Euclidean geometry no longer apply:

Triangles: In hyperbolic geometry, the sum of the angles of a triangle is always less than 180 degrees. The difference between 180 degrees and the sum of the angles is proportional to the area of the triangle.

Parallel Lines: Through any given point not on a line, there are infinitely many lines that do not intersect the original line. These lines are known as hyperbolic parallels, and they diverge as they extend.

Circles: The circumference and area of a circle in hyperbolic geometry grow exponentially with the radius, unlike in Euclidean geometry, where they grow quadratically.

Hyperbolic geometry has fascinating implications for our understanding of space and shape. For example, in a hyperbolic plane, it is possible to fit an infinite number of triangles with a finite total area, a property that has no counterpart in Euclidean geometry.

ELLIPTIC GEOMETRY

Elliptic geometry, on the other hand, describes a space that is curved positively, like a sphere. This geometry also departs significantly from Euclidean principles:

Triangles: In elliptic geometry, the sum of the angles of a triangle is always greater than 180 degrees. The amount by which the sum exceeds 180 degrees is proportional to the area of the triangle.

Parallel Lines: In elliptic geometry, there are no true parallel lines. All lines eventually intersect. This is similar to the behavior of great circles on a sphere, which always intersect at two points.

Circles: The circumference and area of a circle in elliptic geometry grow differently than in Euclidean geometry, reflecting the positive curvature of the space.

Elliptic geometry has applications in understanding the geometry of the Earth and other spherical objects. It also provides insights into the properties of spaces with positive curvature, which are important in fields such as cosmology and general relativity.

THE IMPACT OF NON-EUCLIDEAN GEOMETRY

The development of non-Euclidean geometries had profound implications for mathematics and our understanding of space. These new geometries showed that the parallel postulate was not a necessary truth but rather a specific property of flat, Euclidean space. By exploring spaces with different curvatures, mathematicians opened up new possibilities for understanding the structure of the universe.

Non-Euclidean geometries also paved the way for further developments in mathematics and physics. The idea that space could be curved was a crucial precursor to the development of Einstein's theory of general relativity, which describes gravity as the curvature of spacetime. Non-Euclidean geometries provided the mathematical framework needed to understand and describe this curvature.

VISUALIZING NON-EUCLIDEAN GEOMETRY

Visualizing non-Euclidean geometries can be challenging, as they differ significantly from our everyday experiences. However, there are several ways to gain intuition about these geometries:

1. **Models of Hyperbolic Space**: One way to visualize hyperbolic geometry is through models such as the Poincaré disk model or the hyperboloid model. These models represent hyperbolic space in a way that preserves its geometric properties, allowing us to explore the behavior of shapes and lines in this curved space.

2. **Spherical Geometry**: Elliptic geometry can be visualized by considering the surface of a sphere. Great circles on a sphere serve as analogs for straight lines, and the properties of triangles and angles on the sphere illustrate the principles of elliptic geometry.

WHY STRAIGHT LINES DON'T EXIST: EXPLORING GEOMETRIC TRUTHS

3. **Tessellations**: Tessellations, or tilings, can also help us understand non-Euclidean geometries. For example, in hyperbolic geometry, it is possible to create regular tessellations with shapes such as triangles or squares that would not be possible in Euclidean space.

These visualizations provide a way to bridge the gap between abstract mathematical concepts and our intuitive understanding of space.

∴

The development of non-Euclidean geometries in the 19th century challenged the long-standing assumptions of Euclidean geometry and fundamentally altered our understanding of space. Mathematicians such as Gauss, Lobachevsky, and Bolyai demonstrated that alternative geometries were not only possible but also consistent and logically sound. Hyperbolic and elliptic geometries revealed that the concept of a straight line is not universal but depends on the underlying geometry of the space.

These discoveries had far-reaching implications for mathematics and physics, laying the groundwork for further advances in our understanding of the universe. By exploring the properties of curved spaces, mathematicians opened new avenues for research and provided the tools needed to describe the curvature of spacetime in general relativity.

As we move forward in this book, we will continue to build on these insights, delving deeper into the implications of curved spaces and the nature of straight lines. By challenging traditional notions of geometry, we can gain a deeper appreciation for the complexity and beauty of the universe and uncover the deeper geometric truths that lie beyond the illusion of straight lines.

WHY STRAIGHT LINES DON'T EXIST: EXPLORING GEOMETRIC TRUTHS

WHY STRAIGHT LINES DON'T EXIST: EXPLORING GEOMETRIC TRUTHS

Albert Einstein's theory of general relativity revolutionized our understanding of space and time. General relativity describes gravity as the curvature of spacetime caused by massive objects. According to this theory, the presence of mass and energy warps the fabric of spacetime, causing paths that would be straight in a flat space to curve.

EINSTEIN'S CURVED SPACE

In the context of general relativity, the closest equivalent to a straight line is a geodesic. A geodesic is the path that a free-falling object follows under the influence of gravity. These paths are curved by the presence of mass and energy, demonstrating that straight lines do not exist in the curved spacetime of our universe.

This chapter explores the implications of general relativity for our understanding of straight lines, highlighting how gravity warps spacetime and affects the trajectories of objects.

THE BIRTH OF GENERAL RELATIVITY

Albert Einstein published his theory of general relativity in 1915, providing a new framework for understanding gravity. Before Einstein, gravity was understood through Isaac Newton's laws, which described gravity as a force between two masses. However, Einstein's theory offered a radically different perspective: gravity is not a force but a curvature of spacetime caused by mass and energy.

WHY STRAIGHT LINES DON'T EXIST: EXPLORING GEOMETRIC TRUTHS

Einstein's famous equation, $E=mc^2$, hinted at the deep relationship between energy and mass, suggesting that mass could affect spacetime itself. This relationship is at the heart of general relativity, where mass and energy determine the curvature of spacetime, and this curvature dictates the motion of objects.

SPACETIME AND CURVATURE

In general relativity, spacetime is a four-dimensional continuum that combines the three dimensions of space with the dimension of time. The presence of mass and energy curves this spacetime, creating what we perceive as gravity.

Imagine placing a heavy ball on a rubber sheet. The ball causes the sheet to curve around it, creating a dip. If you roll a smaller ball across the sheet, it will follow a curved path around the dip, much like how planets orbit stars. This analogy helps visualize how massive objects like stars and planets curve spacetime, affecting the paths of other objects.

GEODESICS: THE NEW STRAIGHT LINES

In this curved spacetime, the concept of a straight line from Euclidean geometry no longer applies. Instead, the closest equivalent to a straight line is a geodesic. A geodesic is the path that an object follows when it is not acted upon by any forces other than gravity. For example, the orbit of a planet around the Sun is a geodesic in the curved spacetime created by the Sun's mass.

WHY STRAIGHT LINES DON'T EXIST: EXPLORING GEOMETRIC TRUTHS

Geodesics are curved paths in spacetime that generalize the idea of straight lines to curved spaces. They represent the shortest path between two points in curved spacetime, analogous to how straight lines represent the shortest path in flat space.

THE WARPING OF SPACETIME BY GRAVITY

The presence of mass and energy warps spacetime in a way that can be quantified by Einstein's field equations. These equations describe how mass and energy influence the curvature of spacetime and, in turn, how this curvature affects the motion of objects.

1. **Massive Objects**: Massive objects like stars and planets create significant curvature in spacetime. This curvature causes nearby objects to follow curved paths. For instance, the Earth's orbit around the Sun is a result of the Sun's massive curvature of spacetime.

2. **Light Bending**: Even light, which has no mass, follows curved paths in the presence of massive objects. This phenomenon, known as gravitational lensing, occurs because spacetime itself is curved, bending the path of light. Gravitational lensing has been observed in the universe, confirming the predictions of general relativity.

EXPERIMENTAL EVIDENCE FOR GENERAL RELATIVITY

WHY STRAIGHT LINES DON'T EXIST: EXPLORING GEOMETRIC TRUTHS

General relativity has been confirmed by numerous experiments and observations. Some key pieces of evidence include:

1. **Mercury's Orbit**: The orbit of Mercury around the Sun precesses in a way that cannot be explained by Newtonian mechanics alone. General relativity accurately predicts this precession, accounting for the curvature of spacetime caused by the Sun's mass.

2. **Gravitational Lensing**: The bending of light by massive objects has been observed in astronomical phenomena. For example, light from distant stars can be bent around galaxies, creating multiple images of the same star. These observations match the predictions of general relativity.

3. **GPS Systems**: The Global Positioning System (GPS) relies on general relativity to provide accurate location data. The satellites that make up the GPS network experience different gravitational forces than objects on Earth's surface, affecting the passage of time. General relativity accounts for these differences, ensuring the precision of GPS technology.

BLACK HOLES: EXTREME CURVATURE OF SPACETIME

One of the most dramatic predictions of general relativity is the existence of black holes, regions of space where the curvature of spacetime becomes so extreme that not even light can escape. Black holes form when massive stars collapse under their own gravity, creating a singularity—a point of infinite density—and an event horizon—the boundary beyond which nothing can return.

WHY STRAIGHT LINES DON'T EXIST: EXPLORING GEOMETRIC TRUTHS

The study of black holes has provided deep insights into the nature of spacetime and gravity. Observations of objects orbiting black holes, the detection of gravitational waves from black hole mergers, and the imaging of black holes' event horizons have all confirmed the extreme predictions of general relativity.

GRAVITATIONAL WAVES: RIPPLES IN SPACETIME

In addition to predicting the curvature of spacetime, general relativity also predicts the existence of gravitational waves—ripples in spacetime caused by the acceleration of massive objects. These waves travel at the speed of light and can be produced by events such as the collision of black holes or neutron stars.

In 2015, the Laser Interferometer Gravitational-Wave Observatory (LIGO) made the first direct detection of gravitational waves, confirming another key prediction of general relativity. These detections have opened a new window into the universe, allowing scientists to observe cosmic events that were previously invisible.

THE IMPLICATIONS FOR STRAIGHT LINES

The concept of straight lines, as understood in Euclidean geometry, does not apply in the curved spacetime of general relativity. Instead, geodesics represent the natural paths that objects follow under the influence of gravity. These geodesics are curved by the presence of mass and energy, demonstrating that what we perceive as straight lines are actually the result of complex interactions in a curved spacetime.

WHY STRAIGHT LINES DON'T EXIST: EXPLORING GEOMETRIC TRUTHS

VISUALIZING CURVED SPACETIME

Visualizing curved spacetime can be challenging, as it involves understanding a four-dimensional continuum. However, there are several ways to gain intuition about these concepts:

1. **Rubber Sheet Analogy**: As mentioned earlier, the rubber sheet analogy helps visualize how massive objects curve spacetime. Placing a heavy object on a rubber sheet creates a dip, and smaller objects follow curved paths around this dip.

2. **Light Cones**: In general relativity, light cones represent the possible paths that light can take from a given event in spacetime. The curvature of spacetime affects the shape and orientation of these light cones, illustrating how gravity influences the paths of light and matter.

3. **Computer Simulations**: Modern computer simulations can model the curvature of spacetime and the motion of objects within it. These simulations provide visual representations of phenomena such as black holes, gravitational waves, and the orbits of planets, helping to bridge the gap between abstract mathematical concepts and our intuitive understanding.

∴

Albert Einstein's theory of general relativity revolutionized our understanding of space and time, describing gravity as the curvature of spacetime caused by mass and energy. In this framework, the concept of straight lines from Euclidean geometry is replaced by geodesics—curved

WHY STRAIGHT LINES DON'T EXIST: EXPLORING GEOMETRIC TRUTHS

paths that objects follow under the influence of gravity. General relativity has been confirmed by numerous experiments and observations, including the precession of Mercury's orbit, gravitational lensing, and the detection of gravitational waves.

The implications of general relativity challenge our traditional notions of geometry, revealing that what we perceive as straight lines are actually the result of complex interactions in a curved spacetime. As we continue to explore the nature of spacetime and gravity, we gain a deeper appreciation for the complexity and beauty of the universe, moving beyond the illusion of straight lines to uncover the true geometric truths that shape our reality.

At the quantum level, the behavior of particles further disrupts the idea of straight lines. Quantum mechanics reveals that particles do not follow precise paths. Instead, their positions and momenta are described by probability distributions. Heisenberg's Uncertainty Principle states that we cannot simultaneously know the exact position and momentum of a particle. Consequently, the concept of a well-defined, straight-line trajectory loses its meaning at microscopic scales. The probabilistic nature of particle paths challenges the classical notion of straight lines, introducing a new layer of complexity to our understanding of geometry.

THE QUANTUM WORLD

THE BIRTH OF QUANTUM MECHANICS

Quantum mechanics emerged in the early 20th century as scientists attempted to explain phenomena that classical physics could not account for. Pioneers such as Max Planck, Niels Bohr, Werner Heisenberg, and Erwin Schrödinger developed a new framework for understanding the behavior of particles at atomic and subatomic scales.

Classical physics, based on the deterministic laws of Newtonian mechanics, described particles as having well-defined positions and velocities, following precise paths through space. However, experiments such as the double-slit experiment revealed that particles like electrons exhibit both particle-like and wave-like behavior, leading to the development of quantum mechanics.

WAVE-PARTICLE DUALITY

One of the fundamental concepts of quantum mechanics is wave-particle duality, which posits that particles can exhibit both wave-like and particle-like properties. This duality was first observed in experiments involving light and later confirmed for particles such as electrons.

In the double-slit experiment, electrons are fired at a barrier with two slits. When one slit is open, the electrons form a pattern on a detector consistent with particle behavior. However, when both slits are open, the electrons create an interference pattern, indicative of wave behavior. This experiment demonstrates that particles do not follow well-defined paths but instead exist as probability waves that interfere with each other.

THE WAVE FUNCTION AND PROBABILITY DISTRIBUTIONS

The wave function, introduced by Erwin Schrödinger, is a mathematical description of the quantum state of a particle. The wave function contains all the information about a particle's position, momentum, and other properties. However, the wave function itself is not directly observable; instead, its square gives the probability distribution of finding the particle in a particular state.

WHY STRAIGHT LINES DON'T EXIST: EXPLORING GEOMETRIC TRUTHS

For example, the wave function of an electron in an atom describes the probability distribution of the electron's position around the nucleus. This distribution is not a single point but a cloud of probabilities, indicating where the electron is likely to be found.

HEISENBERG'S UNCERTAINTY PRINCIPLE

Werner Heisenberg formulated the Uncertainty Principle, which states that we cannot simultaneously know the exact position and momentum of a particle. Mathematically, the Uncertainty Principle is expressed as:

$$\Delta x \cdot \Delta p \geq \frac{h}{4\pi}$$

where Δx is the uncertainty in position, Δp is the uncertainty in momentum, and h is Planck's constant.

This principle implies that the more precisely we know a particle's position, the less precisely we can know its momentum, and vice versa. As a result, the concept of a well-defined, straight-line trajectory loses its meaning at the quantum level. Particles do not travel along precise paths but are described by probability distributions that reflect their inherent uncertainty.

THE BREAKDOWN OF CLASSICAL TRAJECTORIES

In classical mechanics, particles follow deterministic paths that can be predicted with great accuracy. However, in quantum mechanics, particles do not have definite trajectories. Instead, their behavior is governed by probabilities and uncertainties.

For instance, consider an electron moving through space. In classical mechanics, we could predict its path based on its initial position and velocity. In quantum mechanics, however, the electron's position and momentum are described by a wave function, which evolves over time according to Schrödinger's equation. The electron does not follow a single path but rather a range of possible paths, each with its own probability.

QUANTUM TUNNELING

One of the most striking consequences of quantum mechanics is the phenomenon of quantum tunneling. In classical physics, a particle must have enough energy to overcome a potential barrier. However, in quantum mechanics, particles can "tunnel" through barriers even when they do not have enough energy to overcome them.

Quantum tunneling occurs because the particle's wave function extends beyond the barrier, allowing a small probability of the particle being found on the other side. This phenomenon has been observed in numerous experiments and has practical applications in technologies such as tunnel diodes and quantum computing.

IMPLICATIONS FOR GEOMETRY AND STRAIGHT LINES

The probabilistic nature of particle paths in quantum mechanics challenges the classical notion of straight lines. At the quantum level, particles do not follow well-defined trajectories but exist as probability clouds. This uncertainty disrupts the concept of a straight line, which relies on the idea of a precise, unambiguous path between two points.

In quantum mechanics, the "path" of a particle is a superposition of all possible paths, each with its own probability amplitude. This idea is encapsulated in Richard Feynman's path integral formulation, which describes the behavior of particles as a sum over all possible histories. In this framework, the concept of a single, well-defined path (or straight line) is replaced by a multitude of possible paths, each contributing to the overall behavior of the particle.

QUANTUM ENTANGLEMENT AND NON-LOCALITY

Another fundamental aspect of quantum mechanics is entanglement, a phenomenon where particles become correlated in such a way that the state of one particle is dependent on the state of another, even across vast distances. This non-locality challenges our classical understanding of space and separateness.

Entangled particles do not have independent, well-defined states. Instead, their properties are linked, such that measuring one particle instantaneously affects the state of the other, regardless of the distance between them. This interconnectedness suggests that our classical notions of space, and by extension straight lines, are incomplete when describing the quantum realm.

WHY STRAIGHT LINES DON'T EXIST: EXPLORING GEOMETRIC TRUTHS

VISUALIZING QUANTUM PATHS

Visualizing quantum paths can be challenging, given the abstract nature of wave functions and probability distributions. However, several approaches can help build intuition:

1. **Probability Clouds**: Visual representations of electron clouds in atoms illustrate the probabilistic nature of particle positions. These clouds show regions where the electron is most likely to be found, highlighting the uncertainty and complexity of quantum paths.

2. **Feynman Diagrams**: Feynman diagrams provide a graphical representation of particle interactions in quantum field theory. These diagrams depict all possible paths and interactions, emphasizing the multiplicity of quantum paths and the breakdown of classical straight lines.

3. **Double-Slit Experiment**: Visualizations and simulations of the double-slit experiment reveal the wave-like interference patterns created by particles. These patterns underscore the departure from classical trajectories and the emergence of probabilistic behavior.

∴

At the quantum level, the behavior of particles fundamentally challenges the classical notion of straight lines. Quantum mechanics reveals that particles do not follow precise paths but are described by probability distributions. Heisenberg's Uncertainty Principle states that we cannot simultaneously know the exact position and momentum of a particle, leading to the breakdown of well-defined trajectories at microscopic scales.

WHY STRAIGHT LINES DON'T EXIST: EXPLORING GEOMETRIC TRUTHS

The probabilistic nature of particle paths, quantum tunneling, and entanglement all disrupt the classical idea of straight lines, introducing new layers of complexity to our understanding of geometry. As we continue to explore the quantum world, we gain a deeper appreciation for the intricacies of reality, moving beyond classical straight lines to uncover the rich, probabilistic tapestry that defines the quantum realm.

Through the lens of quantum mechanics, we see that straight lines are not fundamental features of nature but rather idealized constructs that simplify the complex, uncertain behavior of particles. This understanding invites us to rethink our classical notions of geometry and embrace the profound insights that quantum mechanics offers about the nature of reality.

WHY STRAIGHT LINES DON'T EXIST: EXPLORING GEOMETRIC TRUTHS

The rotation and revolution of celestial bodies add another layer to the understanding of why straight lines do not exist in reality. The Earth rotates on its axis approximately once every 24 hours and revolves around the Sun once every 365.25 days. This means that any point on the Earth's surface is constantly moving in a circular path due to rotation and an elliptical path due to revolution.

THE SPINNING UNIVERSE

Additionally, the Sun itself orbits the center of the Milky Way galaxy, moving at an average speed of about 220 kilometers per second (490,000 miles per hour). The entire solar system, including the Earth, is moving through the galaxy. These combined motions mean that any attempt to draw a straight line is influenced by multiple layers of movement, each adding its own curvature to the perceived straightness.

This chapter explores the dynamic movements of celestial bodies and how they influence our perception of straight lines, introducing the concept of spiraling through space.

EARTH'S ROTATION AND REVOLUTION

The Earth experiences two primary motions: rotation and revolution. These motions have profound effects on our perception of straight lines and the overall geometry of our environment.

WHY STRAIGHT LINES DON'T EXIST: EXPLORING GEOMETRIC TRUTHS

ROTATION

The Earth rotates on its axis once every 24 hours. This rotation causes the alternation of day and night and creates a centrifugal force that affects the shape of the Earth. Due to this rotation, the Earth is not a perfect sphere but an oblate spheroid, slightly flattened at the poles and bulging at the equator.

The rotational speed at the equator is approximately 1670 kilometers per hour (1040 miles per hour). As we move towards the poles, the rotational speed decreases, becoming zero at the poles. This rotation means that any point on the Earth's surface is constantly moving in a circular path. Consequently, any line we draw on the surface of the Earth is subject to this continuous motion.

REVOLUTION

The Earth revolves around the Sun once every 365.25 days in an elliptical orbit. This revolution is responsible for the changing seasons as the tilt of the Earth's axis causes different parts of the Earth to receive varying amounts of sunlight throughout the year.

The average speed of the Earth's revolution around the Sun is about 30 kilometers per second (67,000 miles per hour). This means that while the Earth is rotating on its axis, it is also moving along an elliptical path around the Sun. Therefore, any line drawn on the Earth's surface is not only influenced by the rotational motion but also by the revolution of the Earth around the Sun.

WHY STRAIGHT LINES DON'T EXIST: EXPLORING GEOMETRIC TRUTHS

THE SUN'S MOTION THROUGH THE MILKY WAY

The Sun, along with the entire solar system, orbits the center of the Milky Way galaxy. This motion adds another layer of complexity to the movement of celestial bodies.

GALACTIC ORBIT

The Sun orbits the center of the Milky Way galaxy at an average speed of about 220 kilometers per second (490,000 miles per hour). It takes approximately 225-250 million years for the Sun to complete one orbit around the galactic center. This period is known as a cosmic year or galactic year.

The path of the Sun around the Milky Way is not a perfect circle but a complex, oscillating trajectory influenced by the gravitational forces of other stars and galactic structures. As the Sun moves through the galaxy, it drags the entire solar system along with it, including the Earth.

THE COMBINED MOTIONS

The combined motions of the Earth's rotation, revolution around the Sun, and the Sun's orbit around the Milky Way mean that any point on the Earth's surface is following a highly complex, spiraling path through space. This path is not a straight line but a series of intertwined curves and spirals influenced by multiple gravitational forces and movements.

WHY STRAIGHT LINES DON'T EXIST: EXPLORING GEOMETRIC TRUTHS

SPIRALING THROUGH SPACE

To visualize the complexity of these combined motions, imagine a spiral within a spiral. The Earth rotates on its axis, creating a daily spiral. At the same time, the Earth revolves around the Sun, adding another layer of spiraling motion. Meanwhile, the Sun, carrying the entire solar system, moves through the galaxy, adding yet another spiral to the overall trajectory.

This intricate, multi-layered spiraling motion means that any line we attempt to draw on the Earth's surface is not fixed or straight in an absolute sense. Instead, it is a segment of a larger, dynamic spiral that constantly changes with the movements of celestial bodies.

THE IMPLICATIONS FOR PERCEIVED STRAIGHT LINES

Understanding the dynamic movements of celestial bodies and their combined effects on our perception of straight lines challenges the traditional notions of geometry. In a universe where everything is in constant motion, the idea of a fixed, straight line becomes untenable.

RELATIVE MOTION

The concept of relative motion is crucial in understanding why straight lines do not exist in reality. Relative motion refers to the motion of an object as observed from a particular frame of reference. In the case of the Earth, our frame of reference is constantly changing due to its rotation, revolution, and the Sun's galactic orbit.

WHY STRAIGHT LINES DON'T EXIST: EXPLORING GEOMETRIC TRUTHS

When we draw a line on the Earth's surface, we perceive it as straight relative to our immediate surroundings. However, this perception is limited to our local frame of reference. In the larger context of the Earth's and the solar system's movements, the line is part of a complex, spiraling trajectory.

THE CURVATURE OF SPACE

The dynamic motions of celestial bodies also interact with the curvature of spacetime described by Einstein's theory of general relativity. Massive objects like the Earth and the Sun curve the spacetime around them, influencing the paths of objects and the lines we perceive.

In the curved spacetime of our universe, geodesics (the closest equivalents to straight lines) are not straight in the Euclidean sense but follow the curvature created by gravitational forces. The paths of objects in this curved spacetime are inherently curved, further challenging the notion of straight lines.

VISUALIZING THE SPINNING UNIVERSE

Visualizing the combined motions of celestial bodies and their effects on perceived straight lines can be challenging. However, several approaches can help build intuition:

1. **Rotating Globes**: Observing the rotation and revolution of globes or models of the Earth and the solar system can illustrate the complex, intertwined motions of celestial bodies.

WHY STRAIGHT LINES DON'T EXIST: EXPLORING GEOMETRIC TRUTHS

2. **Computer Simulations**: Modern computer simulations can model the dynamic motions of celestial bodies and the resulting spiraling paths. These simulations provide visual representations of the intricate trajectories followed by objects in the universe.

3. **Spiral Models**: Creating physical models of spirals within spirals can help visualize the combined effects of rotation, revolution, and galactic motion on perceived straight lines.

∴

The rotation and revolution of celestial bodies, along with the Sun's motion through the Milky Way, add multiple layers of complexity to our understanding of straight lines. These combined motions create intricate, spiraling paths that challenge the traditional notions of fixed, straight lines.

In a universe where everything is in constant motion, the idea of a straight line becomes an illusion limited to our local frame of reference. The dynamic movements of celestial bodies and the curvature of spacetime further emphasize that straight lines do not exist in reality but are constructs of human perception.

As we continue to explore the nature of the universe and its dynamic motions, we gain a deeper appreciation for the complexity and beauty of the cosmos. By understanding the spiraling trajectories that define our world, we move beyond the illusion of straight lines to uncover the true geometric truths that shape our reality.

WHY STRAIGHT LINES DON'T EXIST: EXPLORING GEOMETRIC TRUTHS

In Indian philosophy, the concepts of Maya (illusion) and Leela (divine play) provide profound insights into the nature of reality. Maya, according to Advaita Vedanta, is the cosmic illusion that veils the true nature of Brahman, the ultimate reality. It creates the appearance of a diverse, tangible world where distinctions and forms seem real but are ultimately illusory. Straight lines, as part of this perceived world, are thus illusions within the grand illusion of Maya.

MAYA AND LEELA - PHILOSOPHICAL PERSPECTIVES

Leela, in Vaishnavism, refers to the divine play of the universe, where the material world and its phenomena are seen as the playful expressions of the divine will. The perceived order, including geometric constructs like straight lines, are manifestations of this divine play, lacking independent reality.

This chapter delves into the philosophical perspectives of Maya and Leela, exploring how these concepts explain the illusory nature of straight lines.

WHY STRAIGHT LINES DON'T EXIST: EXPLORING GEOMETRIC TRUTHS

THE CONCEPT OF MAYA IN ADVAITA VEDANTA

UNDERSTANDING MAYA

Maya is a fundamental concept in Advaita Vedanta, a school of Hindu philosophy that emphasizes non-dualism. According to Advaita Vedanta, Brahman is the ultimate, unchanging reality, and everything else is an illusion created by Maya. This illusion includes the material world and all its forms, distinctions, and phenomena.

Maya is not merely an illusion in the sense of being false or unreal; it is a powerful force that shapes our perception of reality. It creates a world that seems tangible and distinct, masking the underlying unity of Brahman. The illusion of separateness and multiplicity arises from Maya, leading us to perceive the world as composed of discrete objects and entities.

THE ROLE OF PERCEPTION

Perception plays a crucial role in the experience of Maya. Our senses and mind are conditioned to perceive and interpret the world in certain ways, creating a framework of understanding that aligns with the illusion of Maya. This framework includes geometric constructs like straight lines, which help us navigate and make sense of the world.

However, these constructs are ultimately part of the illusory nature of Maya. They are tools of perception that simplify and organize our experiences but do not reflect the true nature of reality. In the context of Maya, straight lines are convenient abstractions that facilitate our interaction with the material world, yet they lack independent existence.

THE UNVEILING OF REALITY

Advaita Vedanta teaches that the realization of Brahman involves the dissolution of Maya. This process is known as Moksha or liberation, where one transcends the illusions of the material world and perceives the true, undivided reality of Brahman. Through practices such as meditation, self-inquiry, and the study of sacred texts, individuals can awaken to the realization that the distinctions and forms created by Maya are illusory.

In this enlightened state, the constructs of straight lines and other geometric forms are seen for what they are: tools of perception within the grand illusion of Maya. The realization of Brahman reveals the underlying unity and interconnectedness of all things, transcending the illusory distinctions created by Maya.

LEELA: THE DIVINE PLAY OF THE UNIVERSE

THE CONCEPT OF LEELA

Leela, in Vaishnavism, refers to the divine play of the universe. It is the idea that the material world and all its phenomena are expressions of the divine will, manifesting as a cosmic play. Unlike Maya, which is often viewed as a veil of illusion, Leela is seen as a joyful and purposeful expression of the divine.

WHY STRAIGHT LINES DON'T EXIST: EXPLORING GEOMETRIC TRUTHS

Leela encompasses the entire universe, including the creation, sustenance, and dissolution of worlds. It is the dynamic, creative aspect of the divine, where the perceived order and structure of the universe are seen as playful manifestations of the divine will.

THE PLAYFUL NATURE OF REALITY

In the context of Leela, the perceived order of the universe, including geometric constructs like straight lines, is understood as part of the divine play. These constructs are not ultimate realities but temporary expressions of the divine, serving specific purposes within the cosmic drama.

The playful nature of Leela suggests that the universe is not bound by rigid structures or fixed forms. Instead, it is fluid and dynamic, constantly evolving and changing. The perceived straight lines and other geometric forms are transient and flexible, reflecting the ever-changing nature of the divine play.

EMBRACING THE PLAY

Vaishnavism encourages devotees to embrace the divine play of Leela with joy and reverence. By recognizing the playful nature of reality, individuals can cultivate a sense of detachment and acceptance, understanding that the material world and its constructs are temporary manifestations of the divine will.

In this perspective, the illusion of straight lines becomes an invitation to participate in the cosmic play with a sense of wonder and appreciation. The awareness of Leela allows individuals to navigate the world with a lighthearted and open-minded approach, embracing the fluidity and impermanence of all forms.

MAYA AND LEELA: COMPLEMENTARY PERSPECTIVES

THE INTERPLAY OF ILLUSION AND PLAY

Maya and Leela are complementary concepts that offer a holistic understanding of reality in Indian philosophy. While Maya emphasizes the illusory nature of the material world, Leela highlights the dynamic and playful aspect of the divine manifestation. Together, they provide a nuanced view of existence, where the perceived order and structure of the universe are both illusory and divine.

The interplay of Maya and Leela reveals that the constructs we rely on, such as straight lines, are not fixed realities but part of a larger, more intricate cosmic design. These constructs serve specific purposes within the illusion and play of the universe, guiding our perceptions and interactions while ultimately pointing to the deeper truths of existence.

TRANSCENDING ILLUSION AND EMBRACING PLAY

The journey of self-realization involves transcending the illusions of Maya and embracing the playful nature of Leela. Through spiritual practices, individuals can awaken to the understanding that the distinctions and forms perceived in the material world are transient and illusory. This awakening leads to a deeper appreciation of the divine play, where the constructs of straight lines and other geometric forms are seen as expressions of the divine will.

By transcending the illusion of straight lines, individuals can cultivate a more expansive and flexible perspective, recognizing the fluidity and interconnectedness of all things. This perspective invites a playful and joyful engagement with the world, embracing the divine play of Leela with gratitude and reverence.

IMPLICATIONS FOR GEOMETRIC CONSTRUCTS

STRAIGHT LINES AS ILLUSORY CONSTRUCTS

From the perspectives of Maya and Leela, straight lines are seen as illusory constructs that facilitate our interaction with the material world. These constructs simplify and organize our perceptions, allowing us to navigate and understand our environment. However, they do not represent the ultimate reality but are temporary expressions within the grand illusion of Maya and the divine play of Leela.

FLUIDITY AND FLEXIBILITY IN GEOMETRY

Recognizing the illusory nature of straight lines invites a shift towards fluidity and flexibility in our understanding of geometry. Rather than adhering to rigid, fixed constructs, we can embrace the dynamic and evolving nature of geometric forms. This shift aligns with the insights of modern physics and

WHY STRAIGHT LINES DON'T EXIST: EXPLORING GEOMETRIC TRUTHS

mathematics, which reveal the complexity and interconnectedness of the universe.

By integrating the philosophical perspectives of Maya and Leela, we can cultivate a more holistic and expansive approach to geometry, appreciating the intricate and playful nature of the cosmic design.

∴

In Indian philosophy, the concepts of Maya and Leela provide profound insights into the nature of reality, revealing the illusory and playful aspects of existence. Maya, as the cosmic illusion, creates the appearance of a diverse and tangible world, including constructs like straight lines. Leela, as the divine play, emphasizes the dynamic and joyful expression of the divine will in the material world.

Together, these perspectives challenge the traditional notions of fixed, rigid geometric constructs, inviting us to recognize the fluidity and interconnectedness of all things. By understanding the illusory nature of straight lines and embracing the playful nature of reality, we can cultivate a deeper appreciation for the complexity and beauty of the universe, moving beyond the limitations of perception to uncover the true geometric truths that shape our reality.

WHY STRAIGHT LINES DON'T EXIST: EXPLORING GEOMETRIC TRUTHS

Jagrat, or the waking state, is one of the four states of consciousness described in the Mandukya Upanishad, an ancient Indian text that delves into the nature of reality and consciousness. In this state, we perceive the world through our senses and engage in daily activities. It is a state dominated by Maya, the cosmic illusion that shapes our perception of reality. In Jagrat, straight lines and other geometric constructs become essential tools for communication and understanding, allowing us to navigate and interact with the world.

JAGRAT - THE WAKING STATE

THE ROLE OF MAYA IN JAGRAT

Maya, as discussed in previous chapters, is the cosmic illusion that veils the true nature of Brahman, the ultimate reality. In the waking state, Maya manifests as the distinct and tangible forms we perceive. This illusion creates a sense of separation and multiplicity, leading us to see the world as composed of discrete objects and entities.

In Jagrat, Maya conditions our senses and mind to interpret the world in specific ways, including the perception of geometric constructs like straight lines. These constructs help us organize and make sense of our experiences, but they do not reflect the ultimate reality. Instead, they are

WHY STRAIGHT LINES DON'T EXIST: EXPLORING GEOMETRIC TRUTHS

part of the illusory framework created by Maya to facilitate our interaction with the material world.

TOOLS OF COMMUNICATION AND UNDERSTANDING

In the waking state, straight lines and other geometric constructs serve as essential tools for communication and understanding. Just as travelers use maps, signs, and schedules in an airport lounge to navigate their journeys, humans use these geometric constructs to make sense of their experiences and interact with one another.

STRAIGHT LINES AS NAVIGATIONAL AIDS

Straight lines are fundamental to our spatial understanding and navigation. They represent the shortest distance between two points, making them invaluable for measuring distances, creating maps, and designing structures. In everyday life, we rely on straight lines to plan routes, construct buildings, and understand the layout of our environment.

For example, architects use straight lines to create blueprints and plans for buildings, ensuring that structures are safe and aesthetically pleasing. Engineers use straight lines to design machinery and infrastructure, providing precision and stability. Navigators use straight lines to plot courses on maps, facilitating efficient travel by land, sea, and air.

WHY STRAIGHT LINES DON'T EXIST: EXPLORING GEOMETRIC TRUTHS

GEOMETRIC CONSTRUCTS IN COMMUNICATION

Geometric constructs, including straight lines, also play a crucial role in communication. They provide a common language for expressing ideas and conveying information. For instance, in mathematics and science, geometric diagrams and graphs use straight lines to represent relationships and data, making complex concepts more accessible and understandable.

In art, straight lines help create perspective and depth, allowing artists to depict three-dimensional space on a two-dimensional surface. This use of geometry enhances our visual communication, enabling us to convey spatial relationships and create visually compelling works.

INHERENT LIMITATIONS OF PERCEPTION

While straight lines and geometric constructs are indispensable tools in the waking state, they are ultimately limited by the inherent constraints of human perception and cognition. Our senses and mind are conditioned by Maya to perceive the world in specific ways, leading us to construct and rely on geometric abstractions that simplify our experiences.

THE ILLUSORY NATURE OF GEOMETRIC CONSTRUCTS

Geometric constructs like straight lines are symbolic representations that help us navigate the complexities of life, but they do not represent the ultimate reality. In the context of Maya, these constructs are part of the grand illusion that shapes our perception. They are useful abstractions that facilitate our interaction with the material world, but they lack independent existence.

The recognition of these limitations invites us to question and re-evaluate our reliance on geometric constructs. By understanding their illusory nature, we can cultivate a more flexible and expansive perspective, acknowledging the fluidity and interconnectedness of all things.

TRANSCENDING THE ILLUSION

Transcending the illusion of Maya involves recognizing and moving beyond the limitations of our perceptual constructs. In the waking state, this process can be initiated through practices such as meditation, self-inquiry, and the study of sacred texts. These practices help us cultivate awareness of the underlying unity of Brahman, revealing the true nature of reality beyond the illusory distinctions created by Maya.

As we deepen our understanding, we can begin to see straight lines and other geometric constructs for what they are: tools of perception within the grand illusion of Maya. This realization allows us to navigate the waking state with greater clarity and insight, embracing the dynamic and interconnected nature of reality.

THE ANALOGY OF THE AIRPORT LOUNGE

An intriguing analogy compares the waking state (Jagrat) to an airport lounge. In this metaphor, the airport lounge is a transient space where travelers from different destinations interact and prepare for their journeys.

WHY STRAIGHT LINES DON'T EXIST: EXPLORING GEOMETRIC TRUTHS

Similarly, the waking state is where individuals engage with the world and each other, navigating the transient, illusory nature of physical reality.

TRANSIENCE AND INTERACTION

The airport lounge is a place of constant movement and interaction. Travelers come and go, preparing for their next journey while sharing the same space temporarily. This transience mirrors the impermanent and ever-changing nature of the waking state, where experiences and interactions are fleeting and dynamic.

In Jagrat, we navigate a world shaped by Maya, interacting with others and our environment based on our perceptions and constructs. These interactions are influenced by the tools of communication and understanding that we rely on, including straight lines and geometric forms.

NAVIGATIONAL TOOLS IN THE LOUNGE

Just as travelers use maps, signs, and schedules to navigate the airport lounge, humans use geometric constructs to navigate the waking state. These tools provide a framework for understanding our environment and interacting with others, facilitating our journey through the transient space of the waking state.

In the airport lounge analogy, the maps and signs represent the straight lines and geometric constructs that help us make sense of our surroundings. They are essential for efficient navigation and communication, yet they are also temporary and context-dependent, reflecting the illusory nature of the waking state.

EMBRACING THE INHERENT LIMITATIONS

Recognizing the inherent limitations of our perceptual constructs invites us to adopt a more flexible and open-minded approach to navigating the waking state. By acknowledging the illusory nature of straight lines and other geometric forms, we can cultivate a deeper appreciation for the complexity and fluidity of reality.

FLEXIBILITY IN PERCEPTION

Flexibility in perception involves being open to the dynamic and evolving nature of our experiences. Instead of rigidly adhering to fixed constructs, we can embrace the fluidity of geometric forms, understanding that they are tools of perception rather than ultimate realities.

This flexibility allows us to adapt to changing circumstances and new insights, fostering a more holistic and expansive perspective. It encourages us to see beyond the apparent solidity of straight lines and geometric constructs, appreciating the underlying interconnectedness of all things.

CULTIVATING AWARENESS

Cultivating awareness of the illusory nature of the waking state is a key aspect of spiritual practice in many philosophical traditions. By developing

WHY STRAIGHT LINES DON'T EXIST: EXPLORING GEOMETRIC TRUTHS

mindfulness and self-awareness, we can recognize the limitations of our perceptual constructs and transcend the illusions created by Maya.

Practices such as meditation, self-inquiry, and the study of sacred texts can help us deepen our awareness and understanding. These practices encourage us to question and re-evaluate our assumptions, leading to a more profound realization of the true nature of reality.

∴

The Jagrat, or waking state, is a realm dominated by the influence of Maya, where straight lines and other geometric constructs serve as essential tools for communication and understanding. These constructs help us navigate and interact with the world, yet they are ultimately part of the grand illusion created by Maya.

An analogy comparing the waking state to an airport lounge highlights the transient and dynamic nature of this state, where geometric constructs function as navigational aids. Recognizing the inherent limitations of our perceptual constructs invites us to adopt a more flexible and open-minded approach to navigating the waking state.

By understanding the illusory nature of straight lines and embracing the fluidity of geometric forms, we can cultivate a deeper appreciation for the complexity and interconnectedness of reality. This awareness allows us to transcend the limitations of perception and uncover the true geometric truths that shape our reality, moving beyond the constructs of the waking state to embrace the deeper insights of existence.

WHY STRAIGHT LINES DON'T EXIST: EXPLORING GEOMETRIC TRUTHS

WHY STRAIGHT LINES DON'T EXIST: EXPLORING GEOMETRIC TRUTHS

By integrating various perspectives from mathematics, science, and metaphysics, we arrive at a unified understanding that straight lines are constructs of the mind rather than features of the physical world. These constructs help us navigate and understand our environment, but they are ultimately idealized abstractions that do not reflect the true nature of space and reality.

INTEGRATING PERSPECTIVES

MATHEMATICS AND IDEALIZATION

Mathematics provides idealized models that help us understand and manipulate our environment. Euclidean geometry, with its straight lines, is one such model. However, non-Euclidean geometries remind us that these models are abstractions, not reflections of the true nature of space.

EUCLIDEAN GEOMETRY

In Euclidean space, straight lines are defined as the shortest distance between two points. This concept is foundational to many mathematical applications and is used extensively in fields such as engineering, architecture, and navigation. The simplicity and practicality of Euclidean

WHY STRAIGHT LINES DON'T EXIST: EXPLORING GEOMETRIC TRUTHS

geometry make it a useful tool for understanding and interacting with the world.

However, Euclidean geometry operates under specific assumptions, such as the flatness of space and the infinite extension of lines, which do not hold true in the actual physical universe. These assumptions simplify the complexities of space, allowing us to create clear and precise models, but they do not account for the curvature of space caused by massive objects or the dynamic nature of the universe.

NON-EUCLIDEAN GEOMETRY

The development of non-Euclidean geometries in the 19th century challenged the traditional notions of Euclidean geometry. Mathematicians such as Carl Friedrich Gauss, Nikolai Lobachevsky, and János Bolyai explored alternative geometries that did not adhere to Euclid's fifth postulate, which states that parallel lines never meet.

Hyperbolic Geometry: In hyperbolic geometry, space is curved negatively, like a saddle. In this geometry, the sum of the angles of a triangle is less than 180 degrees, and parallel lines can diverge. Hyperbolic geometry reveals that the concept of a straight line is not universal but depends on the underlying geometry of the space.

Elliptic Geometry: In elliptic geometry, space is curved positively, like a sphere. In this geometry, the sum of the angles of a triangle is more than 180 degrees, and there are no true parallel lines; all lines eventually intersect. Elliptic geometry further challenges the notion of straight lines,

WHY STRAIGHT LINES DON'T EXIST: EXPLORING GEOMETRIC TRUTHS

showing that in curved spaces, geodesics (the closest equivalent to straight lines) are curved paths that reflect the underlying geometry of the space.

These non-Euclidean geometries expand our understanding of space and demonstrate that the concept of a straight line is an idealization rather than a fundamental feature of reality.

SCIENCE AND EMPIRICAL REALITY

Science, through the theories of general relativity and quantum mechanics, provides empirical evidence that further challenges the notion of straight lines as fundamental features of reality. These theories reveal the complex, dynamic nature of the universe and the limitations of classical geometric constructs.

GENERAL RELATIVITY

Albert Einstein's theory of general relativity revolutionized our understanding of space and time by describing gravity as the curvature of spacetime caused by massive objects. In this framework, the concept of a straight line from Euclidean geometry is replaced by geodesics—curved paths that objects follow under the influence of gravity.

Curvature of Spacetime: Massive objects like stars and planets warp the fabric of spacetime, causing paths that would be straight in a flat space to curve. This curvature affects the motion of objects and the paths of light,

WHY STRAIGHT LINES DON'T EXIST: EXPLORING GEOMETRIC TRUTHS

demonstrating that straight lines do not exist in the curved spacetime of our universe.

Experimental Evidence: General relativity has been confirmed by numerous experiments and observations, including the precession of Mercury's orbit, gravitational lensing, and the detection of gravitational waves. These findings support the idea that spacetime is curved and that the concept of a straight line is not applicable in this context.

QUANTUM MECHANICS

At the quantum level, the behavior of particles further disrupts the idea of straight lines. Quantum mechanics reveals that particles do not follow precise paths but are described by probability distributions.

Heisenberg's Uncertainty Principle: This principle states that we cannot simultaneously know the exact position and momentum of a particle. Consequently, the concept of a well-defined, straight-line trajectory loses its meaning at microscopic scales. The probabilistic nature of particle paths challenges the classical notion of straight lines, introducing a new layer of complexity to our understanding of geometry.

Wave-Particle Duality: Particles exhibit both wave-like and particle-like behavior, and their paths are not fixed but exist as superpositions of all possible paths. This duality further challenges the classical idea of straight lines and highlights the inherent uncertainty and complexity of the quantum world.

WHY STRAIGHT LINES DON'T EXIST: EXPLORING GEOMETRIC TRUTHS

METAPHYSICS AND ULTIMATE REALITY

Metaphysical insights from Indian philosophy, particularly the concepts of Maya and Leela, provide a deeper context for understanding the illusory nature of straight lines and other geometric constructs.

MAYA AND THE ILLUSION OF DISTINCTIONS

According to Advaita Vedanta, Maya is the cosmic illusion that creates the appearance of a diverse, tangible world. This illusion shapes our perception of reality, leading us to see the world as composed of distinct objects and entities. Straight lines, as part of this perceived world, are thus illusions within the grand illusion of Maya.

Tools of Perception: Geometric constructs like straight lines are tools of perception that help us navigate and understand the material world. However, they do not reflect the ultimate reality but are temporary expressions within the illusion of Maya.

Transcending Maya: The realization of Brahman, the ultimate reality, involves transcending the illusions of Maya and recognizing the underlying unity and interconnectedness of all things. This realization reveals that the distinctions and forms created by Maya, including straight lines, are illusory.

LEELA AND THE DIVINE PLAY

In Vaishnavism, Leela refers to the divine play of the universe, where the material world and its phenomena are seen as playful expressions of the divine will. The perceived order, including geometric constructs like straight lines, is part of this divine play, lacking independent reality.

Dynamic and Playful Nature of Reality: Leela emphasizes the fluid and dynamic nature of the universe, where forms and structures are temporary manifestations of the divine will. Straight lines and other geometric constructs are part of this cosmic play, serving specific purposes within the material world.

Embracing Leela: By recognizing the playful nature of reality, individuals can cultivate a sense of detachment and acceptance, understanding that the material world and its constructs are transient and flexible. This perspective invites a joyful and open-minded engagement with the world, embracing the divine play of Leela.

THE UNIFIED UNDERSTANDING

By integrating these perspectives from mathematics, science, and metaphysics, we arrive at a unified understanding that straight lines are constructs of the mind rather than features of the physical world. These constructs are useful for navigating and understanding our environment but do not reflect the true nature of space and reality.

THE ROLE OF HUMAN PERCEPTION

Human perception plays a crucial role in shaping our understanding of geometric constructs. Our senses and mind are conditioned to interpret the world in specific ways, leading us to construct and rely on idealized

WHY STRAIGHT LINES DON'T EXIST: EXPLORING GEOMETRIC TRUTHS

abstractions like straight lines. These constructs simplify and organize our experiences, facilitating communication and interaction, but they are ultimately limited by the constraints of perception.

MOVING BEYOND CLASSICAL CONSTRUCTS

Recognizing the limitations of classical geometric constructs invites us to move beyond rigid and fixed notions of space. By embracing the insights from non-Euclidean geometries, general relativity, quantum mechanics, and metaphysical philosophies, we can cultivate a more flexible and expansive understanding of geometry.

Fluidity and Interconnectedness: The unified understanding emphasizes the fluidity and interconnectedness of all things, challenging the traditional notion of fixed, straight lines. This perspective aligns with the dynamic and evolving nature of the universe, encouraging us to appreciate the complexity and beauty of reality.

Holistic Approach: Integrating these perspectives provides a holistic approach to understanding geometry, where idealized constructs are seen as useful tools rather than ultimate realities. This approach fosters a deeper appreciation for the intricate and interconnected nature of the cosmos, inviting us to explore the true geometric truths that shape our existence.

∴

By integrating the perspectives from mathematics, science, and metaphysics, we uncover a unified understanding that straight lines are constructs of the mind rather than features of the physical world. These

WHY STRAIGHT LINES DON'T EXIST: EXPLORING GEOMETRIC TRUTHS

constructs help us navigate and understand our environment but are ultimately idealized abstractions that do not reflect the true nature of space and reality.

Euclidean geometry, with its straight lines, provides a useful model for many practical applications, but non-Euclidean geometries reveal that these models are abstractions. General relativity and quantum mechanics further challenge the notion of straight lines, demonstrating the complex and dynamic nature of the universe. Metaphysical philosophies, particularly the concepts of Maya and Leela, offer deeper insights into the illusory and playful nature of geometric constructs.

By embracing these integrated perspectives, we can cultivate a more flexible and expansive understanding of geometry, recognizing the fluidity and interconnectedness of all things. This unified understanding invites us to move beyond the limitations of perception and uncover the true geometric truths that shape our reality, exploring the profound mysteries that lie beyond the constructs of the mind.

WHY STRAIGHT LINES DON'T EXIST: EXPLORING GEOMETRIC TRUTHS

Understanding that straight lines are constructs of perception rather than features of reality has practical implications for various fields. This realization invites us to rethink and innovate in science and engineering, philosophy and metaphysics, and art and design. By embracing the complexity and fluidity of the universe, we can develop more accurate models, deepen our understanding of reality, and create more dynamic and reflective works of art.

PRACTICAL IMPLICATIONS

SCIENCE AND ENGINEERING

Straight lines have been fundamental tools in science and engineering, providing simplicity and precision in measurements, designs, and models. However, acknowledging their limitations can lead to more accurate and comprehensive approaches that account for the curvature of space and the complexities of the natural world.

RETHINKING MODELS AND DESIGNS

Incorporating the understanding that straight lines are idealized constructs rather than absolute realities can improve the accuracy and functionality of scientific models and engineering designs.

WHY STRAIGHT LINES DON'T EXIST: EXPLORING GEOMETRIC TRUTHS

Geodesic Paths in Space Exploration: In space exploration, considering the curvature of spacetime is crucial for planning efficient trajectories for spacecraft. By using geodesics, the equivalent of straight lines in curved spacetime, scientists can optimize flight paths, conserve fuel, and enhance mission success.

Architectural Innovations: In architecture, recognizing the limitations of Euclidean geometry can inspire innovative designs that better reflect the natural environment. Structures that incorporate curves and organic forms can achieve greater harmony with their surroundings and improve structural integrity by distributing stress more evenly.

ENHANCING TECHNOLOGICAL APPLICATIONS

Technological advancements can also benefit from integrating the understanding of curved space and non-linear dynamics.

GPS and Relativity: The Global Positioning System (GPS) relies on general relativity to provide accurate location data. The satellites in the GPS network experience different gravitational forces than objects on Earth's surface, affecting the passage of time. By accounting for these relativistic effects, the system ensures precise navigation and timing.

Quantum Computing: Quantum mechanics, which challenges the classical notion of straight lines, underpins the development of quantum computing. Quantum computers leverage the probabilistic nature of particles and their superposition states to perform complex calculations at unprecedented speeds, promising revolutionary advancements in computing power and efficiency.

WHY STRAIGHT LINES DON'T EXIST: EXPLORING GEOMETRIC TRUTHS

PHILOSOPHY AND METAPHYSICS

Recognizing the illusory nature of straight lines can deepen our understanding of reality and our place within it. This shift in perspective encourages us to move beyond the limitations of perceived forms to embrace the underlying unity and interconnectedness of all things.

EXPANDING CONSCIOUSNESS

The realization that straight lines are constructs of perception invites us to expand our consciousness and explore the deeper truths of existence.

Non-Dualistic Awareness: In Advaita Vedanta, the recognition of Maya and the illusory nature of distinctions leads to the realization of Brahman, the ultimate reality. This non-dualistic awareness transcends the limitations of perceived forms and reveals the underlying unity of all things, fostering a deeper sense of interconnectedness and harmony.

Holistic Understanding: Embracing the concept of Leela, the divine play, encourages a holistic understanding of reality. By recognizing the dynamic and playful nature of the universe, we can cultivate a sense of detachment and acceptance, appreciating the transient and interconnected aspects of existence.

PRACTICAL PHILOSOPHICAL APPLICATIONS

This expanded understanding can also influence practical philosophical applications in daily life.

Mindfulness and Presence: Practicing mindfulness and presence can help individuals transcend the illusions of perceived forms and connect with the deeper reality. By cultivating awareness of the present moment and the interconnectedness of all things, individuals can foster a greater sense of peace, clarity, and purpose.

Ethical and Sustainable Living: Recognizing the interconnectedness of all things can inspire more ethical and sustainable living practices. By understanding the impact of our actions on the broader web of life, we can make more conscious choices that promote the well-being of all beings and the planet.

ART AND DESIGN

Artists and designers can use the concept of curved space and non-linear forms to create works that reflect the true nature of reality, challenging conventional notions of geometry and space.

INNOVATIVE ARTISTIC EXPRESSIONS

Embracing the complexity and fluidity of the universe can inspire innovative artistic expressions that capture the dynamic nature of reality.

Organic and Fluid Forms: Artists can explore organic and fluid forms that move beyond rigid, straight lines to convey the dynamic and interconnected nature of the universe. These forms can evoke a sense of movement, growth, and transformation, inviting viewers to engage with the deeper aspects of reality.

WHY STRAIGHT LINES DON'T EXIST: EXPLORING GEOMETRIC TRUTHS

Interdisciplinary Approaches: By integrating insights from mathematics, science, and metaphysics, artists can create interdisciplinary works that bridge the gap between art and other fields. These works can offer new perspectives on reality and challenge viewers to reconsider their assumptions about space and form.

FUNCTIONAL AND AESTHETIC DESIGN

Designers can also apply these principles to create functional and aesthetic designs that reflect the true nature of reality.

Curved and Ergonomic Designs: In product and industrial design, incorporating curved and ergonomic forms can enhance functionality and user experience. By considering the natural movements and interactions of users, designers can create more intuitive and comfortable products.

Biomimicry and Sustainable Design: Biomimicry, the practice of emulating nature's designs and processes, can lead to more sustainable and efficient solutions. By studying the curves and forms found in nature, designers can develop innovative products and systems that harmonize with the environment and promote sustainability.

∴

Understanding that straight lines are constructs of perception rather than features of reality has profound practical implications for science and engineering, philosophy and metaphysics, and art and design. By acknowledging the limitations of classical geometric constructs and embracing the complexity and fluidity of the universe, we can develop more

WHY STRAIGHT LINES DON'T EXIST: EXPLORING GEOMETRIC TRUTHS

accurate models, deepen our understanding of reality, and create more dynamic and reflective works of art.

In science and engineering, incorporating the understanding of curved space and non-linear dynamics can improve the accuracy and functionality of models and designs, leading to technological advancements and architectural innovations. In philosophy and metaphysics, recognizing the illusory nature of straight lines can expand our consciousness and foster a deeper sense of interconnectedness and harmony. In art and design, embracing curved and organic forms can inspire innovative expressions that capture the dynamic nature of reality and promote more sustainable and ergonomic solutions.

By integrating these perspectives, we can cultivate a more holistic and expansive approach to understanding and interacting with the world. This unified understanding invites us to move beyond the limitations of perception and uncover the true geometric truths that shape our reality, exploring the profound mysteries that lie beyond the constructs of the mind.

WHY STRAIGHT LINES DON'T EXIST: EXPLORING GEOMETRIC TRUTHS

To further explore the illusion of straight lines and deepen your understanding of the underlying geometric truths, consider engaging in the following thought experiments and exercises. These activities are designed to challenge your perceptions and provide practical insights into the concepts discussed in previous chapters.

THOUGHT EXPERIMENTS AND EXERCISES

1. DRAW A LINE IN MOTION

Objective: Understand how motion affects the perceived straightness of a line and relate this to the motion of celestial bodies.

Instructions:

- Take a piece of paper and draw a straight line while standing still. Observe the line and note its straightness.

- Now, walk in a circle and attempt to draw another straight line on the paper. Compare this line with the one drawn while standing still.

WHY STRAIGHT LINES DON'T EXIST: EXPLORING GEOMETRIC TRUTHS

- Reflect on how your motion affected the perceived straightness of the line. Consider how this exercise relates to the rotation and revolution of celestial bodies, which are constantly in motion.

Reflection:

- The line drawn while walking in a circle is likely to be less straight than the one drawn while standing still. This illustrates how motion affects our ability to create and perceive straight lines.

- Apply this understanding to the motion of celestial bodies, such as the Earth's rotation and revolution. Recognize that any line drawn on a moving object is influenced by its motion, reinforcing the idea that straight lines are idealized constructs rather than features of reality.

2. VISUALIZE GEODESICS

Objective: Visualize geodesics on a spherical surface and understand their relationship to curved space in general relativity.

Instructions:

- Obtain a globe or any spherical object.

- Use a marker or string to draw the shortest path between two points on the surface of the sphere. These paths are known as great circles.

- Observe how these paths are curved relative to the surface of the sphere.

WHY STRAIGHT LINES DON'T EXIST: EXPLORING GEOMETRIC TRUTHS

- Reflect on how these curved paths represent the closest equivalent to straight lines in a spherical geometry and how this concept applies to the curvature of space in general relativity.

Reflection:

- Great circles on a sphere are curved but represent the shortest distance between two points on the sphere's surface. This demonstrates that straight lines in Euclidean geometry are not the shortest paths in curved spaces.

- In the context of general relativity, geodesics are the equivalent of straight lines in curved spacetime. This exercise helps visualize how massive objects curve spacetime and how objects follow these curved paths under the influence of gravity.

3. SIMULATE CURVED SPACE

Objective: Create a model to visualize the warping of space by gravity and understand how massive objects affect the trajectories of other objects.

Instructions:

- Use a flexible sheet or piece of fabric to create a simple model of curved space. Stretch the fabric taut and secure it at the edges.

- Place objects of different masses (e.g., balls of varying sizes) on the sheet to simulate the warping of space by gravity. Observe how the sheet curves around the heavier objects.

WHY STRAIGHT LINES DON'T EXIST: EXPLORING GEOMETRIC TRUTHS

- Roll a small ball along the surface of the sheet and observe its curved path as it moves around the heavier objects.

- Reflect on how this model helps visualize the curvature of spacetime and how massive objects influence the paths of other objects.

Reflection:

- The flexible sheet model demonstrates how massive objects cause the fabric of space to curve, creating a visual representation of the warping of spacetime described by general relativity.

- The curved paths of the small ball illustrate how objects follow geodesics in curved spacetime, further reinforcing the concept that straight lines are not features of curved spaces.

4. EXPLORE QUANTUM PATHS

Objective: Simulate the probabilistic nature of particle paths at the quantum level and understand the breakdown of well-defined trajectories.

Instructions:

- On a piece of paper, mark two points representing the starting and ending positions of a particle.

WHY STRAIGHT LINES DON'T EXIST: EXPLORING GEOMETRIC TRUTHS

- Draw multiple possible trajectories for the particle between these two points, considering the uncertainty in its position and momentum. Use dashed or wavy lines to represent the probabilistic nature of these paths.

- Reflect on how the concept of a well-defined straight line breaks down at the quantum scale, where particles exhibit both wave-like and particle-like behavior.

Reflection:

- The multiple trajectories illustrate the probabilistic nature of particle paths in quantum mechanics, where particles do not follow single, well-defined paths.

- This exercise highlights the limitations of classical notions of straight lines and emphasizes the complexity and uncertainty inherent in the quantum world.

∴

Through these thought experiments and exercises, we explore the illusion of straight lines and gain practical insights into the underlying geometric truths. By challenging our perceptions and engaging with these concepts in a hands-on manner, we deepen our understanding of the dynamic, curved, and interconnected nature of reality.

These exercises invite us to move beyond the limitations of classical geometric constructs and embrace a more flexible and expansive perspective. By integrating insights from mathematics, science, and

WHY STRAIGHT LINES DON'T EXIST: EXPLORING GEOMETRIC TRUTHS

metaphysics, we cultivate a holistic understanding of geometry and space, appreciating the profound mysteries that shape our existence.

As we continue to explore and reflect on these concepts, we are reminded that straight lines are constructs of perception rather than features of reality. This realization encourages us to question and re-evaluate our assumptions, leading to a deeper appreciation of the intricate and interconnected nature of the universe. Through this journey, we uncover the true geometric truths that lie beyond the constructs of the mind, embracing the complexity and beauty of the cosmos.

WHY STRAIGHT LINES DON'T EXIST: EXPLORING GEOMETRIC TRUTHS

Straight lines, while fundamental to our understanding and navigation of the world, are ultimately tools of perception, part of the grand illusions of Maya and Leela. Mathematics, science, and metaphysics all converge on this realization, each offering unique insights into why straight lines do not exist in reality but are constructs of the mind.

CONCLUSION: EMBRACING CURVED REALITIES

By exploring the interplay between these disciplines, we gain a deeper appreciation for the complexity and beauty of the universe. We recognize that our perceptions are shaped by underlying principles that transcend simple geometric constructs, inviting us to explore the true, boundless nature of reality beyond the illusions of form and structure. The motion of celestial bodies and the dynamic nature of the universe further emphasize that straight lines are not features of the physical world but convenient abstractions for understanding and navigating our perceived reality.

FINAL THOUGHTS

Through the integration of mathematical idealization, scientific empirical reality, and metaphysical insights, we can see that straight lines are symbolic representations necessary for communication and navigation within the waking state (Jagrat). The Jagrat state, dominated by Maya and

WHY STRAIGHT LINES DON'T EXIST: EXPLORING GEOMETRIC TRUTHS

Leela, provides the framework for interaction and understanding within the confines of human perception and cognition.

Mathematics and Idealization: Euclidean geometry, with its straight lines, serves as a useful model for many practical applications, despite its limitations. Non-Euclidean geometries remind us that these models are abstractions, reflecting the curvature and complexity of real space.

Science and Empirical Reality: General relativity and quantum mechanics challenge the notion of straight lines, showing that spacetime is curved and particle paths are probabilistic. The empirical evidence supports the idea that straight lines are not features of reality but constructs of the mind.

Metaphysics and Ultimate Reality: Insights from Indian philosophy, particularly the concepts of Maya and Leela, suggest that our perceptions, including geometric constructs, are part of a grand illusion. The waking state (Jagrat) is where we navigate this illusion, using straight lines and other tools to make sense of our experiences.

Understanding that straight lines are constructs of perception rather than features of reality invites us to embrace the true, boundless nature of the universe. It encourages a shift in perspective, moving beyond the limitations of perceived forms to recognize the underlying unity and interconnectedness of all things. This journey of exploration and understanding enriches our perspective, guiding us toward a greater appreciation of the profound mysteries that lie beyond the illusions of straight lines and geometric constructs.

EMBRACING A NEW PARADIGM

As we integrate these perspectives, we are invited to embrace a new paradigm that transcends the limitations of classical geometry and rigid constructs. This paradigm shift is not just theoretical but has practical implications across various fields, from science and engineering to philosophy and art. By acknowledging the true nature of reality, we open ourselves to innovative ideas and creative solutions that reflect the dynamic, interconnected fabric of the cosmos.

PRACTICAL IMPLICATIONS

SCIENCE AND ENGINEERING

In science and engineering, the understanding that straight lines are idealized constructs encourages the development of more accurate models and designs that account for the curvature of space and the complexities of the natural world. This leads to advancements in fields such as space exploration, architecture, and technology, where embracing curved paths and non-linear dynamics can result in more efficient and harmonious solutions.

PHILOSOPHY AND METAPHYSICS

Philosophically, recognizing the illusory nature of straight lines deepens our understanding of reality and our place within it. It fosters a sense of unity and interconnectedness, encouraging us to move beyond the limitations of perceived forms and embrace the holistic nature of existence. This shift in perspective can lead to a more profound sense of purpose and meaning in our lives.

ART AND DESIGN

Artists and designers can draw inspiration from the concept of curved space and non-linear forms, creating works that reflect the true nature of reality. This approach challenges conventional notions of geometry and space, resulting in innovative and dynamic artistic expressions that resonate with the fluidity and interconnectedness of the universe.

CONTINUING THE JOURNEY

The exploration of curved realities and the recognition that straight lines are constructs of perception rather than features of reality is an ongoing journey. It invites continuous inquiry, reflection, and adaptation as we deepen our understanding of the universe and our place within it. By integrating insights from mathematics, science, and metaphysics, we cultivate a more comprehensive and enlightened perspective that transcends traditional boundaries and embraces the infinite possibilities of the cosmos.

As we move forward, let us remain open to the wonders of the universe, embracing the complexity and beauty that lie beyond the constructs of the mind. Through this journey, we uncover the profound mysteries that shape our reality, guided by a deeper appreciation for the interconnectedness and fluidity of all things.

WHY STRAIGHT LINES DON'T EXIST: EXPLORING GEOMETRIC TRUTHS

Dr. Sunayana Pandé, affectionately known as Dr. Sun, is a distinguished metaphysician, naturopathic doctor, and transformational coach. With a robust academic background in psychology, neuroscience, and religious studies, Dr. Sun offers a unique blend of scientific and spiritual insights. She is the founder of a temple dedicated to serving non-binary individuals, transgender persons, and drag performers, promoting inclusivity and spiritual growth.

ABOUT THE AUTHOR

Dr. Sun's prolific writing includes several influential books:

- LIFE IN THE BLISS LANE: A GUIDE TO WELLNESS, SELF-LOVE, AND JOY
- YOU MAKE ME SICK: VIRTUE SIGNALING & NARCISSISTIC ABUSE
- HIDDEN BRANCHES IN THE FAMILY TREE: NAVIGATING THE NPE EXPERIENCE
- BEYOND BINARY: AN EXPLORATION IN GENDER AND SEXUALITY
- PHARMAJUANA: GUIDE TO CANNABIS FOR CANCER
- VASUDHAIVA KUTUMBAKAM: THE LIMITLESS POWER OF DESI ROOTS
- SAIL BEYOND TRAUMA: KAPPAL OTTI TRAUMA THERAPY (COMING SOON)

WHY STRAIGHT LINES DON'T EXIST: EXPLORING GEOMETRIC TRUTHS

Her work spans various domains, including wellness, gender and sexuality, trauma therapy, and the therapeutic use of cannabis. Dr. Sun's holistic approach integrates metaphysical principles with practical guidance, helping individuals achieve wellness, self-love, and joy.

To learn more about Dr. Sunayana Pandé and her work, visit her website at http://www.LifeInTheBlissLane.com.

www.ingramcontent.com/pod-product-compliance
Lightning Source LLC
Chambersburg PA
CBHW050233230526
45470CB00005B/1938